Contents

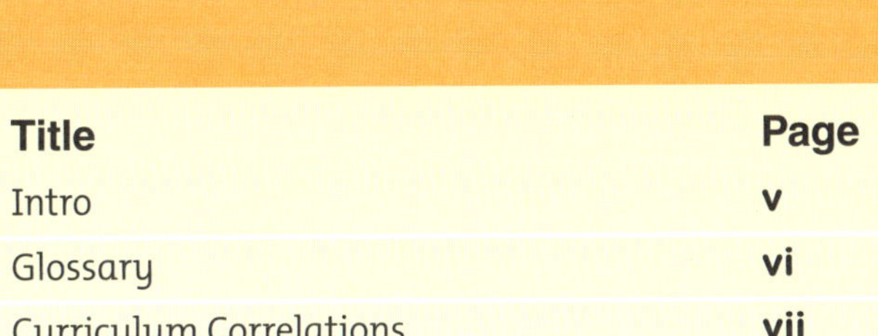

TARGETING SCIENCE FOUNDATION © PASCAL PRESS ISBN: 9781925726497

Chemical Sciences AC9SFU04

Chemical Sciences AC9SFU04

Targeting Science Foundation
Copyright © 2024 Pascal Press

ISBN: 9781925726497

Published by Pascal Press
PO Box 250
Glebe NSW 2037
www.pascalpress.com.au
contact@pascalpress.com.au

This edition of *Skill Sharpeners: Science* is published by arrangement with Evan-Moor Corporation, USA.
For sale in Australia and New Zealand.

Cover Design: Janice Bowles
Authors: Guadalupe Lopez, Lisa Vitarisi Mathews
Publisher: Lynn Dickinson
Editor Australian edition: Stella Tarakson
Typesetter: Stacey Grainger
Illustrator: Paul Lennon, *www.dreamstime.com.au*
Cover design: Janice Bowles

Printed in China by 1010 International Ltd.

Introduction

Welcome to your Foundation *Targeting Science* activity book! It is packed with interesting and exciting activities to help you understand and enjoy science at home or school.

Targeting Science has been written to support the Australian Primary Science Curriculum Version 9.0 and is divided between:

- Biological Sciences
- Physical Sciences
- Chemical Sciences

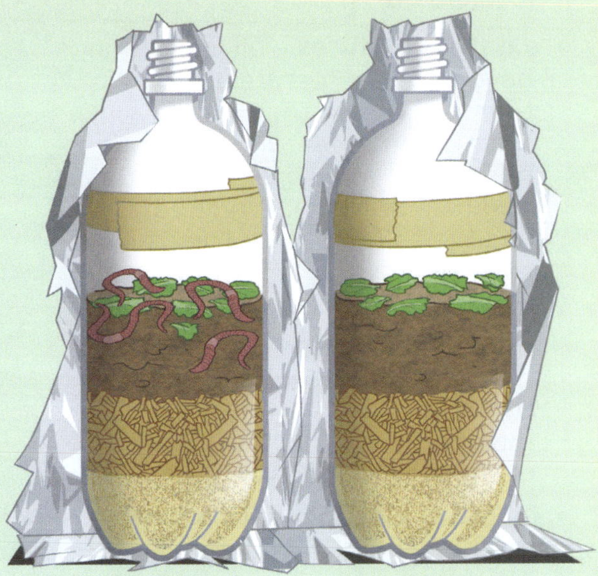

The Science Understanding and Inquiry Skills elements (including Science as a Human Endeavour) are embedded.

Hands-on activities provide opportunities to bring the science concepts to life. The exercises develop process skills such as observing, collecting, recording and organising information and making models of scientific events that happen in the natural world. All activities use easily sourced, inexpensive items.

Topics are introduced with explanations plus images to provide background information. Some lessons include a QR code where you can access a video for further explanations. You will be challenged to match, sort, label, sequence, analyse and answer questions with regular vocabulary practise puzzles throughout the book to help familiarise you with scientific terms. See the Glossary on the next page for some terms you may not already know. Answers are included at the end of the book.

FREE Teaching Guide.

This QR code links to a downloadable PDF of a Teaching Guide to support the material in this student workbook. The guide contains:

- Teaching plans and checklists.
- Graphic organisers.
- Material request forms (to send home for parents).
- Background information about each major topic as well as extension activities.

Use this QR code to access the FREE Teaching Guide.

Front cover – have you looked at the front cover? It depicts what life was like before a significant scientific discovery that led to the development of new technology. What is the difference between Science and Technology? The following quote sums it up nicely.

Science is the process of acquiring knowledge of natural phenomenon along with various reasons. Technology is the application of Science to the solution of problems. (Sismondo, 2018)

Glossary

- **external features** – all living things have external features that help them survive in their habitats. Your external features are your nose, ears, mouth, arms, legs, eyes and hair. Some external features change to help living things to survive.
- **fabric** – cloth or other material produced by weaving together cotton, nylon, wool, silk or other threads. Fabrics are used for making things such as clothes, curtains and sheets.
- **flexible** - capable of bending easily without breaking.
- **flowers** – the part of a plant that blossoms. Flowers produce the seeds that can become new plants. Most plants, including many trees, grow some kind of flower.

- **glass** – a non-living solid material that is usually transparent or translucent as well as hard, brittle and strong against natural elements.
- **material** – a substance with particular qualities or that is used for specific purposes.
- **metal** - a material that, when freshly prepared, polished or fractured, shows a shiny appearance and is a good conductor of electricity and heat.
- **objects** – an object is designed and made from materials.
- **observe** – to watch carefully with attention to detail and behaviour.
- **plastic** – a group of malleable, lightweight materials that can be produced from natural resources or synthetically created.
- **predict** – to suggest a result or outcome based on prior knowledge, experiences, observations and research.
- **properties** – attributes of an object or material, normally used to describe attributes common to a group.
- **senses** – hearing, sight, smell, touch and taste.
- **sort** – finding things that are the same, alike or different and grouping them according to their similarities or differences.
- **stems** – the main structures of a plant that supports the leaves and flowers.
- **texture** – how something feels when touched. It could be smooth or rough, warm or cold.

SAFETY
All of the investigations in this book are designed for kids to do safely in the home or classroom. However, adult supervision is recommended.

TARGETING SCIENCE FOUNDATION © PASCAL PRESS ISBN: 9781925726497

Foundation Australian Science Curriculum Correlations

ACARA Code	Content description	Strand	Sub strand	Pages
AC9SFU01	Observe external features of plants and animals and describe ways they can be grouped based on these features	Science Understanding	Biological Sciences	2-37
AC9SFU02	Describe how objects move and how factors including their size, shape or material influence their movement	Science Understanding	Physical Sciences	38-67
AC9SFU03	Recognise that objects can be composed of different materials and describe the observable properties of those materials	Science Understanding	Chemical Sciences	68-94
AC9SFH01	Explore the ways people make and use observations and questions to learn about the natural world	Science as Human Endeavour	Use and Influence of Science	24, 33, 34, 42, 43, 56, 90
AC9SFI01	Pose questions and make predictions based on experiences	Science Inquiry	Questioning and Predicting	4, 5, 35, 49, 52, 53, 58, 59, 60
AC9SFI02	Engage in investigations safely and make observations using their senses	Science Inquiry	Planning and Conducting	22, 23, 32, 45, 49, 51, 57, 75, 85
AC9SFI03	Represent observations in provided templates and identify patterns with guidance	Science Inquiry	Processing, Modelling and Analysing	8, 12, 22, 32, 36, 40, 46, 49, 53, 69, 72, 75
AC9SFI04	Compare observations with predictions with guidance	Science Inquiry	Evaluating	23, 32, 42, 54, 59, 70, 75
AC9SFI05	Share questions, predictions, observations and ideas with others	Science Inquiry	Communicating	2, 8, 32, 33, 36, 40, 41, 43, 46, 54, 72, 79, 84, 86

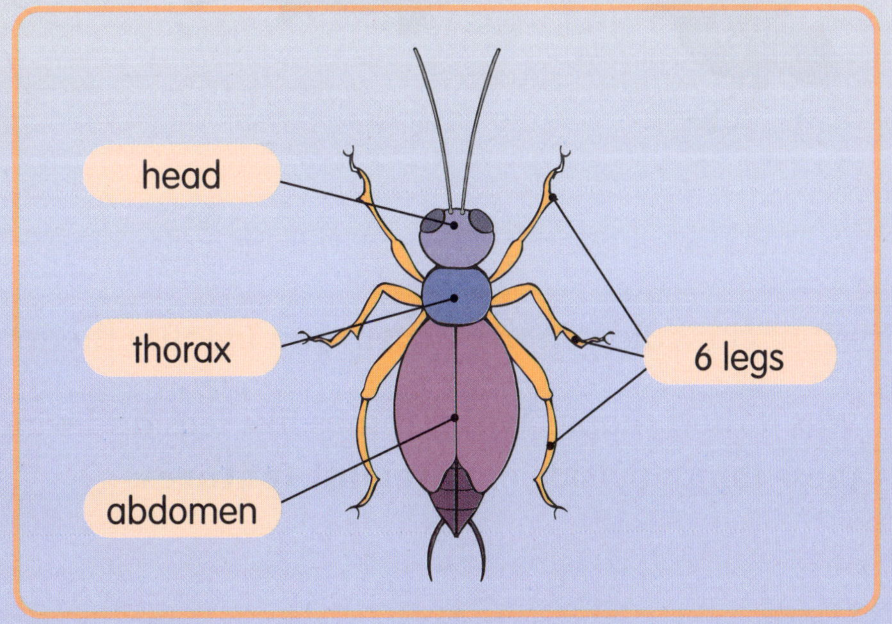

Observing External Features

Scientists, like us, understand the world around them by **observing** and **asking questions**. In this science unit we will learn about how we identify animals and plants based on what they look like. We can then group animals and plants to help us understand them and take care of them.

You are already good at observing. Can you identify what the following pictures are?

What clues did you use to help you identify them?

Was it their shape? Was it things like legs and wings? We call these things outside or **external features**.

External Features of Animals

Through **observing** and **asking questions** you may know that animals come in many different shapes, sizes and colours, but they are all still **animals.**

Look at this picture. Point to any you have seen. Can you name any of them?

What group do these belong to? Animals or plants?

Yes they are all animals! Did you know that people belong to the animal group also?

What **external features** help you know they are animals?

Put a line under all the animals in the picture that you know have 4 legs.

External Features

Animal or Plant?

One of the main ways we can identify if something is an animal or a plant is the shape of its **external features**. Body parts like legs, tails, wings, mouths and ears give animals a shape we can observe. These are the features that help them move around. Parts like leaves, stems and branches give plants a shape we can observe.

Put a circle around all the **animals** in the picture below. Can you identify any of them?

Did you remember that people are animals too?

If you were a scientist, what animal would you like to observe?

What would be a question you would like to ask about this animal?

TARGETING SCIENCE FOUNDATION © PASCAL PRESS ISBN: 9781925726497

External features of plants

Like scientists, you may understand that plants also come in many different shapes, sizes and colours, but they are still plants.

Look at this picture. Point to the ones you have seen. Can you name any of them?

Do they all belong to the plant group?

Yes they are all plants!

What features help you know they are plants?

Put a dot on the plants you think you could climb if they were real.

First Nations Australians' Knowledge of External Features

First Nations Australians are strongly connected to the animals and plants of Australia. They learn how to recognise different plants and animals from when they are children. They love and respect them and to show this they often use animals and plants in their stories, music, dancing and art.

Look at the pictures below. Can you see how the dancers are copying the external features of animals in their dancing? Draw a line from the dancer to an animal they may represent.

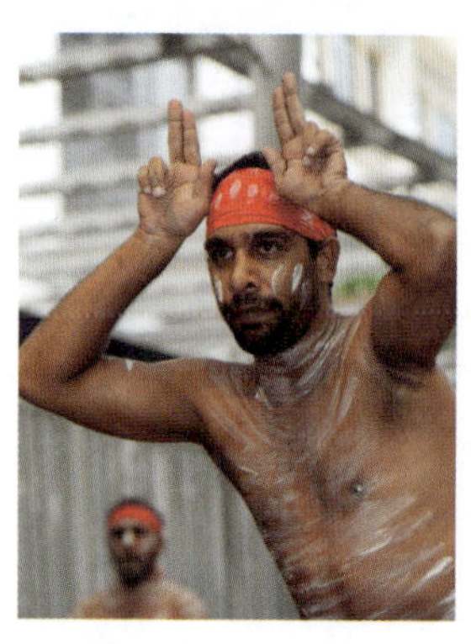

TARGETING SCIENCE FOUNDATION © PASCAL PRESS ISBN: 9781925726497

Using Knowledge of External Features

First Nations Australians tell stories and create art that often includes animals, plants and other features of the land and sea. They observe closely so they show the animals' external features clearly. Draw a line from the art to the animal it shows us.

Grouping by External Features

For this activity you will need a variety of animal models or pictures.

Can you find some toy animals to use for a sorting game? Or perhaps you could use pictures. Observe the features of the different animals closely and think of ways you can group them.

Some ideas are listed below.

Two group ideas:

- Animals with wings and those without wings.
- Animals with legs and those without legs.
- Animals that have fins for swimming and those without.
- Animals bigger than me and animals smaller than me.

Now try your own ways of grouping the animals and share your ideas.

Next look at your own body. Name some of the features that make us people.

How are we like a bird? How are we different?

How are we like a chimpanzee? How are we different?

TARGETING SCIENCE FOUNDATION © PASCAL PRESS ISBN: 9781925726497

Sing this science chant to your child to the tune of "Baa, Baa Black Sheep".

Meow, meow, black cat.
Do you like your **fur**?
Yes, sir. My fur
Is soft and warm.

Peep, peep, white bird.
Do you like your **feathers**?
Yes! They keep me
warm and dry
In cold and rainy weather.

  ISBN: 9781925726497

Sss, sss, slinky snake.
Do you like your **scales**?
My scales protect me
On the rough
and rocky trails.

Knock, knock, turtle.
Do you like your **shell**?
I can tuck my body in.
My shell will
hide me well.

Talk with Your Child

Explain to your child that animals have different coverings that protect them. Tell your child that fur, feathers, scales, and shells are animals' coverings. Help your child name animals that have these coverings.

TARGETING SCIENCE FOUNDATION © PASCAL PRESS ISBN: 9781925726497

An animal's body covering is another external feature

Look at the picture. Read the word.
Colour **yes** ☺. Colour **no** ☹.

scales

shell

feathers

fur

Animal Coverings

Look at the animal's body coverings – fur, feathers, skin, scales and shells.

Draw an **X** on the one that does not belong.

5 What do you notice about the coverings on all of the birds?

TARGETING SCIENCE FOUNDATION © PASCAL PRESS ISBN: 9781925726497

Draw a line to match.

How might the feathers or fur help these animals?

Features and Movement

https://clickv.ie/w/msgx

Use this QR code to access a video on this topic.

Read this science rhyme to your child.

Legs, flippers, wings,
and claws.
Fins, tails, hooves,
and paws.

TARGETING SCIENCE FOUNDATION © PASCAL PRESS ISBN: 9781925726497

In air or water,
land or sky,
watch the animals
moving by.

The body features of an animal tell us a lot about how it can move.

Talk with Your Child
Explain to your child that animals' body parts help them move. Together with your child, examine each animal, pointing to body parts and how each animal can move using them.

Draw a line to match.

TARGETING SCIENCE FOUNDATION © PASCAL PRESS ISBN: 9781925726497

Read the word. Then draw the missing part.

Draw a line to match.

How do the parts help the animal?

TARGETING SCIENCE FOUNDATION © PASCAL PRESS ISBN: 9781925726497

Some animals and plants, or their parts, are so small that scientists have to use special equipment to help them make observations. You may have heard about or even used magnifying glasses and microscopes. These are tools that help us look at things we can't see with our own eyes when they are very small. Like a fly's eye!

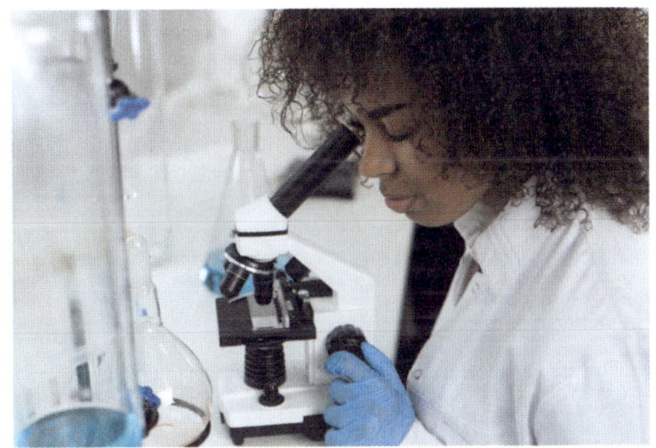

Using a magnifying glass can help us see wonderful things. Using a microscope we can see even more detail!

Here is a fly through a magnifying glass and then through a microscope!

magnifying glass

microscope

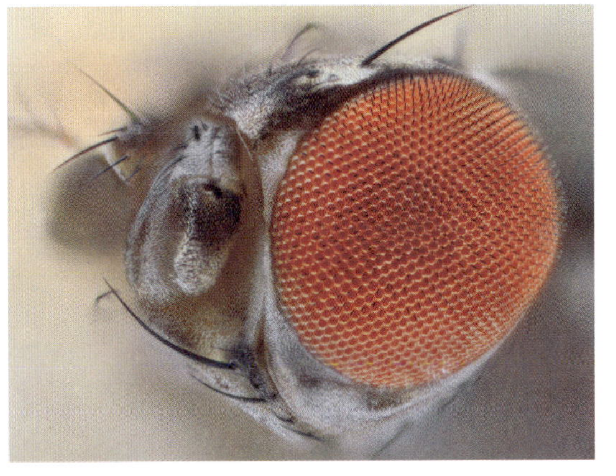

TARGETING SCIENCE FOUNDATION © PASCAL PRESS ISBN: 9781925726497

Observing Like a Scientist

Here is a flower.

This is looking through a magnifying glass...

And this is looking through a microscope...

Try to visit a museum or science centre for the opportunity to look into a microscope. Digital microscopes are also easily sourced.

TARGETING SCIENCE FOUNDATION © PASCAL PRESS ISBN: 9781925726497

Insects

Insects are everywhere.

There are many different kinds of insects.

Like all animals, insects have body parts that allow us to identify them. All insects have 3 body parts and 6 legs. Scientists create drawings with labels to teach us about things.

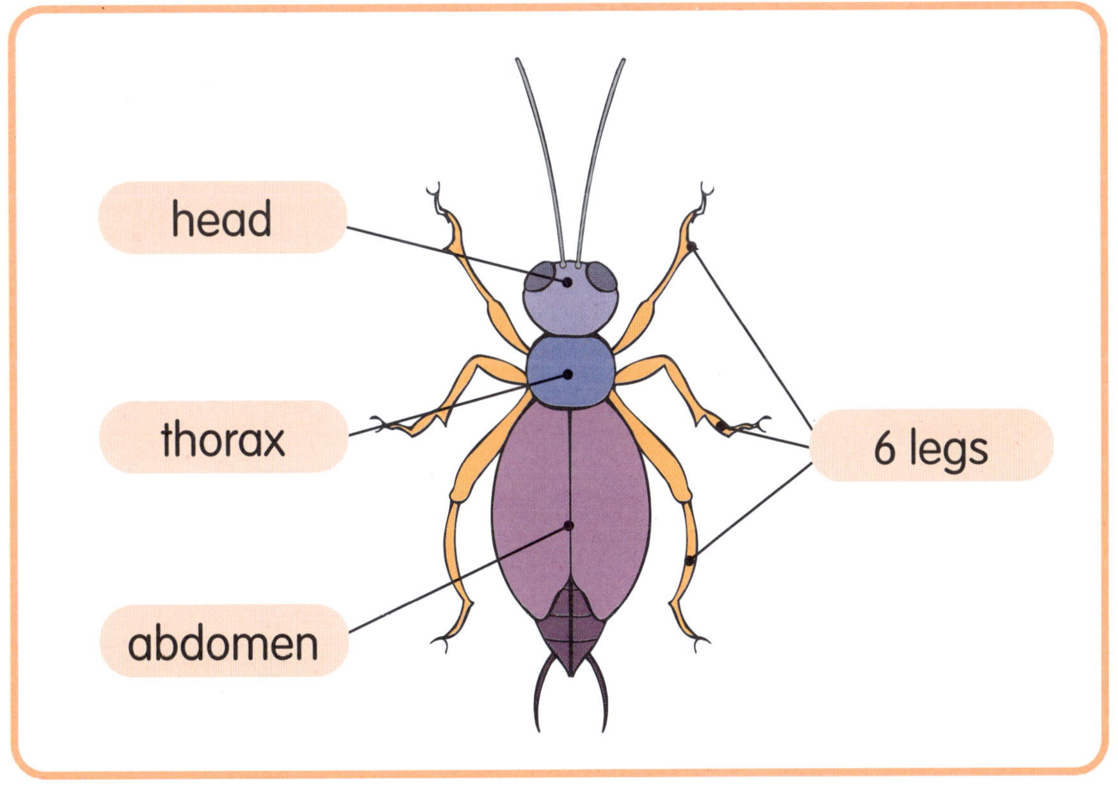

Investigating Insects

Some insects can bite and sting to protect themselves. Scientists, like us, need to observe and respect insects safely. How might we do this? Remember their body parts and go outside to look for insects. Make a ✔ for each insect type you find.

How many of these insect types did you observe?
Did you observe any other types?

TARGETING SCIENCE FOUNDATION © PASCAL PRESS ISBN: 9781925726497

Observing External Features of Plants

Just like trees, smaller plants have parts that we can observe and identify. We know they belong to the plant group because of these features. We can draw a picture to show these parts and add labels to show the different features.

https://clickv.ie/w/uzgx

Use this QR code to access a video on this topic.

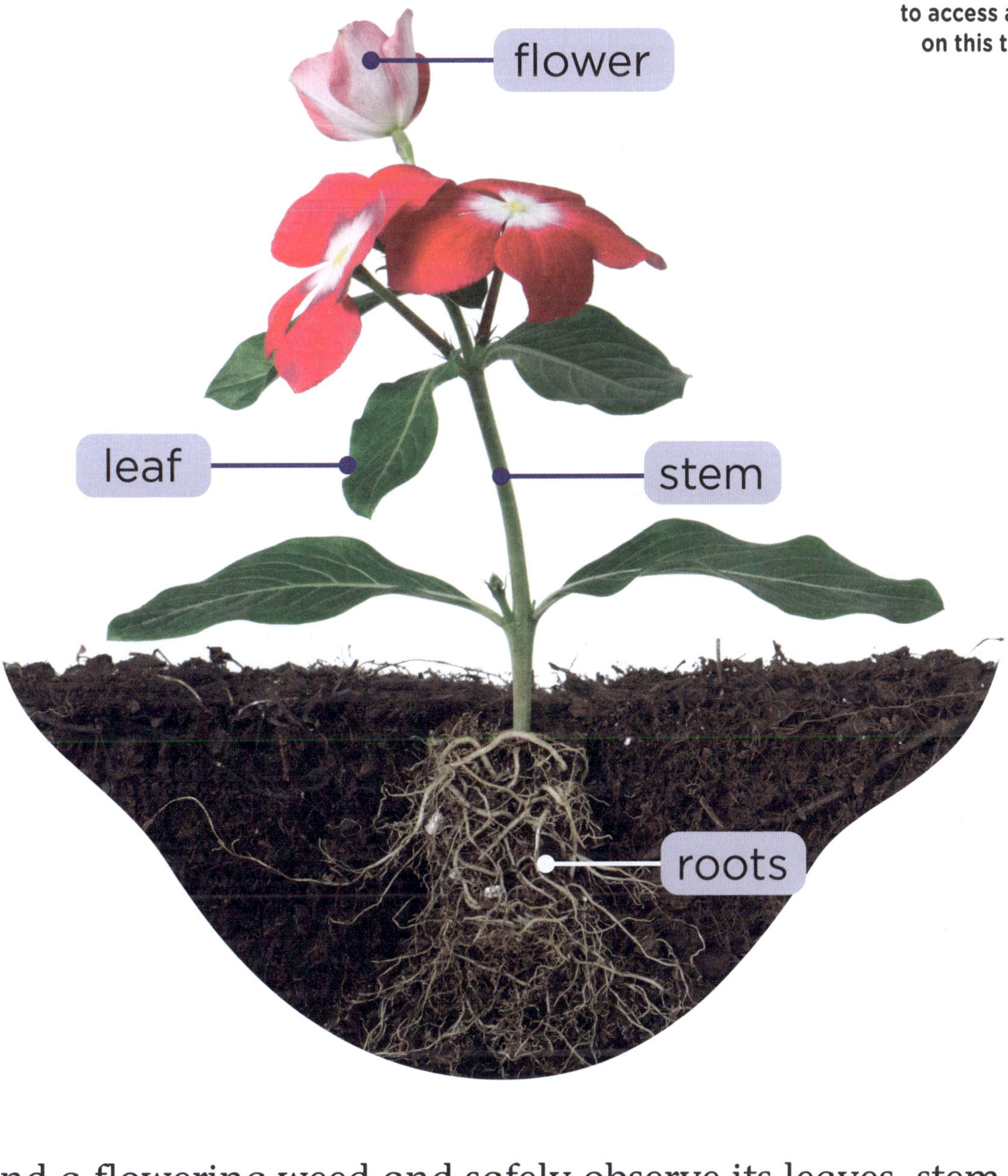

flower

leaf

stem

roots

Find a flowering weed and safely observe its leaves, stem and flowers. Guess what is below the ground. Compare what you see with your guess. Pull it up to observe how the roots help it get water and nutrients, or vitamins, from the soil. Observe how roots also help to hold it in the ground (shelter).

Observing External Features of Plants

Trees are very important in our world. They clean the air we breathe, and they give food and homes for many animals including people. Scientists have learned there are many different types of trees. They know this from observing what trees look like and where they live.

Like animals, trees have parts that we can identify.

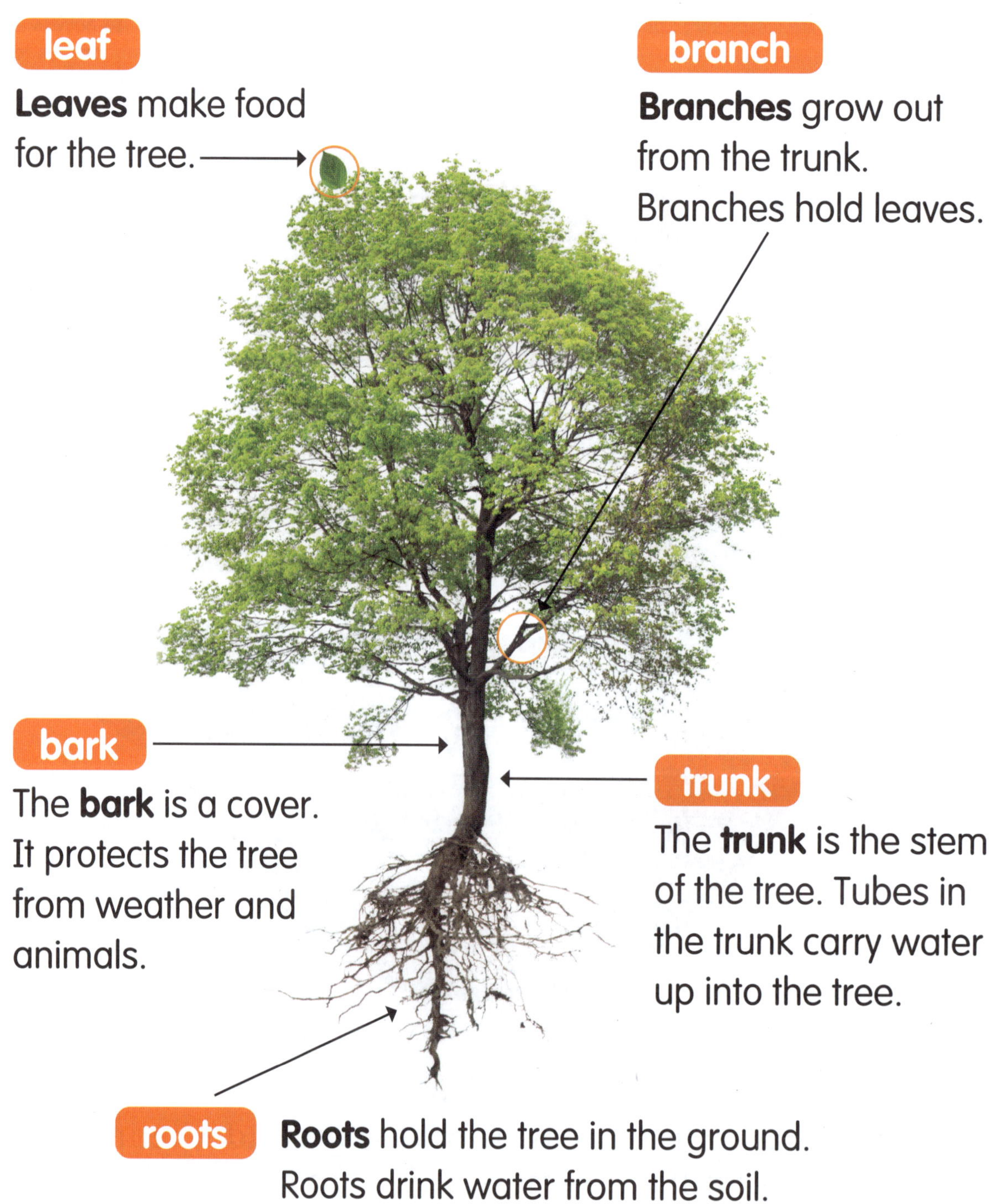

Go outside and observe an old tree and a young tree like a scientist. What features helped you choose? What do you notice?

TARGETING SCIENCE FOUNDATION © PASCAL PRESS ISBN: 9781925726497

Draw the missing parts.

2

3

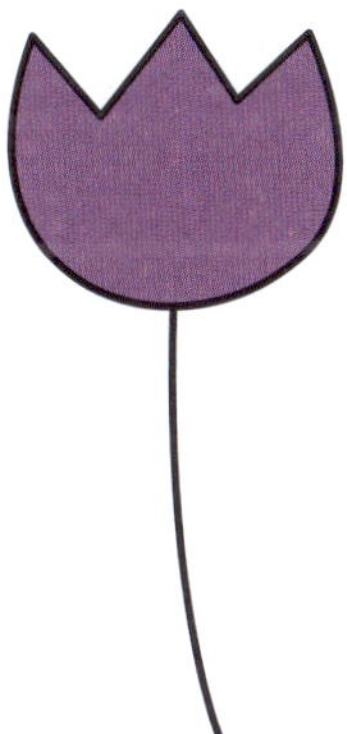

We Eat Plants

Some plants even have parts we can eat.
Read this science rhyme.

Oats, peas, beans,
and pumpkin **seeds**.
These are plant parts
that we need.

Cabbage, spinach,
lettuce, kale.
Leaves of plants
are dark or pale.

Many **fruits**
grow on trees.
Peaches, cherries,
all of these.

 TARGETING SCIENCE FOUNDATION © PASCAL PRESS ISBN: 9781925726497

Apples, oranges,
purple plums.
I like fruits…
Yum, yum, yum!

Celery and
asparagus,
Stems of plants
Are good for us!

Talk with Your Child

Talk with your child about the fruits and vegetables your family eats. Discuss which plant part each of these fruits and vegetables are. Help your child identify his or her favourite plant parts to eat.

We Eat Plant Parts

Many plants have parts we can eat. We can eat the leaves, stems and seeds of some, the roots and fruit of others and sometimes the flowers. Scientists help farmers (and us) grow the healthiest plants they can so we have delicious, healthy food to eat. Draw a line to match the growing plants to the types of foods they give us.

leaves

roots

fruit

seeds

External Features

TARGETING SCIENCE FOUNDATION © PASCAL PRESS ISBN: 9781925726497

Here are some **flowers** we can eat.

Here are some **stems** we can eat.

Look at the picture below. Draw a line from each word to something that matches in the picture.

When you are next helping with cooking or eating dinner, observe different fruits and vegetables and identify what parts of plants they might be.

TARGETING SCIENCE FOUNDATION © PASCAL PRESS ISBN: 9781925726497

Fruit, Root, Seed or Leaf?

Draw a line to match.

TARGETING SCIENCE FOUNDATION © PASCAL PRESS ISBN: 9781925726497

Draw an **X** on the plant part that does not belong.

2

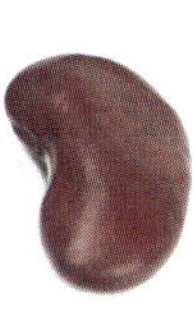

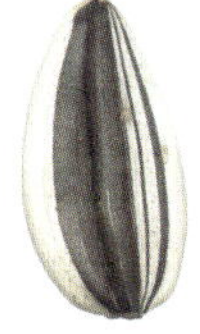

3

Grow a Bean Plant

Plant your own bean seeds and observe what happens.

1. Scrunch a wet paper towel and put it in a clear plastic cup or glass.
   Slide two bean seeds between the towel and the side of the cup.

2. Place the cup in a sunny spot. Water a little when dry. What do you think will happen?

3. After a few days draw what you observe.

4. Plant the seeds with roots in some soil. Water and observe. What do you think will happen if we stop watering it?

External Features

TARGETING SCIENCE FOUNDATION © PASCAL PRESS ISBN: 9781925726497

Using our Science Observation

People ask questions and use science observations of plants and animals in many ways.

Vets - to keep our pets healthy

Zookeepers - to care for animals in the zoo

First Nations Australians - to care for Country

National Park Rangers - to protect nature

Farmers - to grow healthy food for us

Scientists are always observing and asking questions so we can better understand and care for our natural world.

What question would you ask a zookeeper if you could talk with one?

First Nations Australians Using Observations

First Nations Australians know the importance of animals and plants for our survival. They can identify plant parts that are safe to eat. They know what parts of plants can be used to make things. In many Aboriginal grouping systems, plants are sorted into those that give wood, like trees, and those that don't give wood, like grasses and vines. First Nations Australians look at the features of the plants to work out which group they belong to and how the plant can help them.

They can identify particular Eucalyptus tree types so they can find branches that have been eaten out by termites to make Didgeridoos.

They identify Pandanus trees by their shape, leaves and where they grow. They use the leaves for weaving baskets to carry things and mats to sit on.

First Nations Peoples can identify the types of trees that give the best wood for making boomerangs.

TARGETING SCIENCE FOUNDATION © PASCAL PRESS ISBN: 9781925726497

First Nation Australians Using Observations

Many First Nations Peoples sort living things into plant or animal groups. Animal life can then further be sorted into groups, such as land animals, ocean animals and animals with wings, based on the observable features of the creature and where it lives.

Put a circle around the animals that you think belong in the **ocean** animal group.

What is a science question you would ask about one of these animals?

____

____

____

Using our Science Knowledge

Ask an adult to help you locate a place outside to sit and make some observations of plants and animals. Draw the features of 1 **animal** and 1 **plant.**

Name the main external features that help you identify each thing you have drawn.

Explain one way they are the same or different to you.

How can you keep safe when observing animals and plants?

What are some ways people use observation and questioning to learn about plants and animals?

TARGETING SCIENCE FOUNDATION © PASCAL PRESS ISBN: 9781925726497

Using our Science Knowledge

Look at the pictures. Pick one and identify the external features of the animal or plant. Now find other pictures with a matching feature. Make 2 more groups with some or all of the pictures based on features. Explain your groups.

External Features

What is an Object?

There are many, many, things in our world which *can be seen, held or touched.* We use lots of them to play and work. We call these **OBJECTS**. *Objects are not alive.*

Circle all the pictures of objects below.

TARGETING SCIENCE FOUNDATION © PASCAL PRESS ISBN: 9781925726497

What can we Observe About Objects?

When we observe an object, we see and feel many things about it. We see the shape, colour, size and what it is made from. We feel how heavy it is, what it's like to touch and if it can be moved or has moving parts. These observations help us know how we might use the objects.

Look at the pictures below and work out what things we can see and feel about objects.

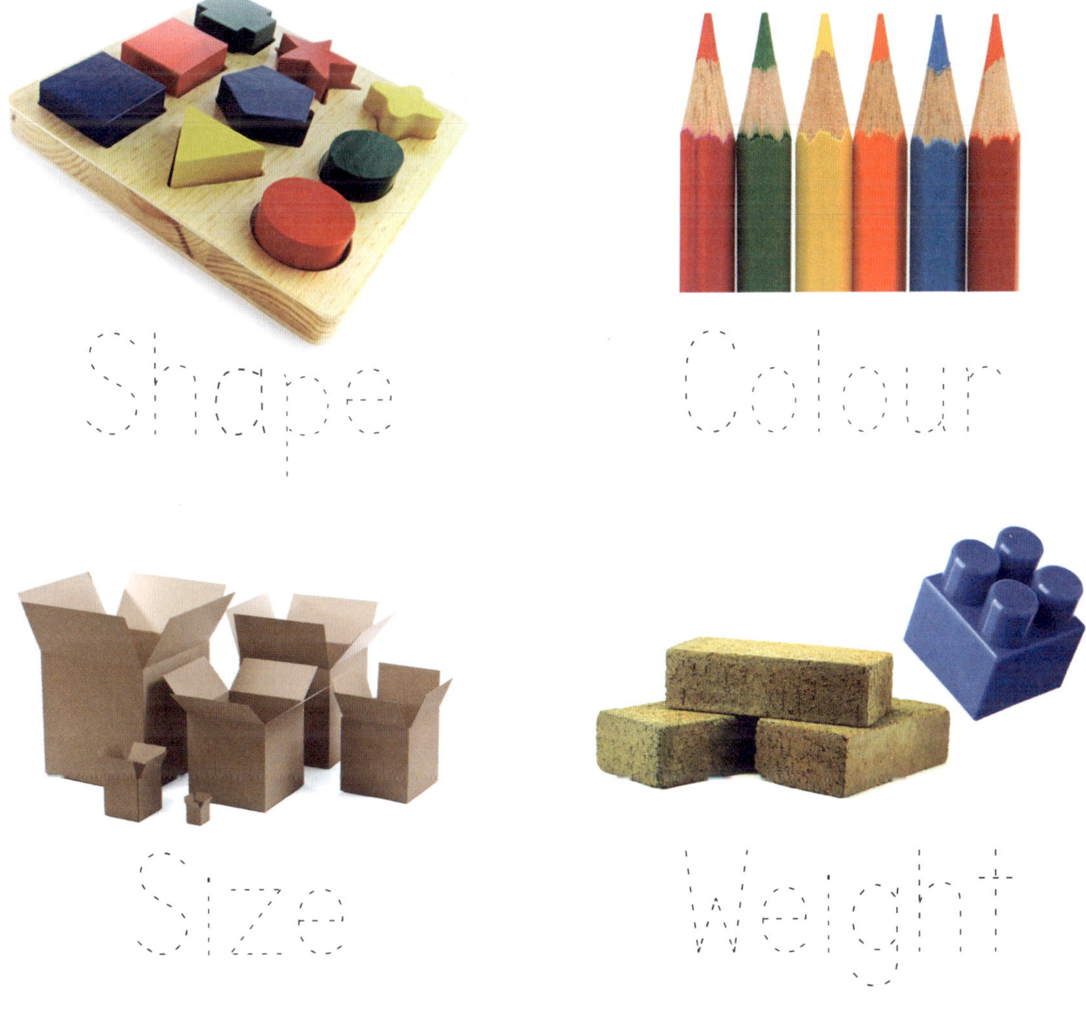

What is it made from?

TARGETING SCIENCE FOUNDATION © PASCAL PRESS ISBN: 9781925726497

Observing Objects

Circle the object picture that is **different** in the row. Why is it different?

Size			
Shape	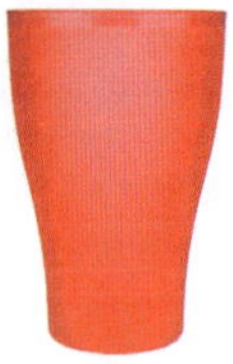		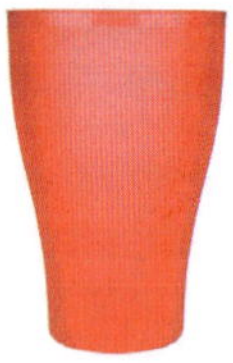
Weight			
Material			
Made to move		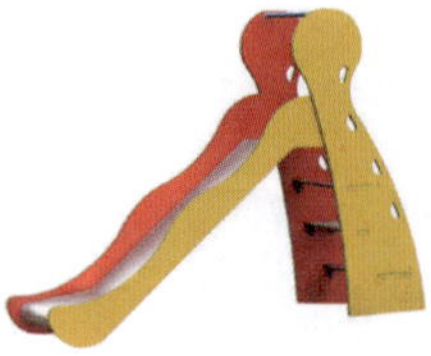	

 TARGETING SCIENCE FOUNDATION © PASCAL PRESS ISBN: 9781925726497

Factors Affecting Object Movement

Some objects are made to be moved and others are not.

Which of these objects could you throw and why?

Of course, you can throw the ball and not the house!

But could you explain why?

 Tick the factors you thought of.

Size	The ball is small	The house is large	
Shape	The ball is round and can be picked up.	The house is joined to the ground and we cannot pick it up.	
Weight	The ball is light.	The house is heavy.	
Material	The ball is made from materials that bounce and roll and won't break.	The house is made from wood and glass and would break if it were thrown.	

These are some of the factors that we observe when we use and move objects.

If the ball was blue, could we still throw it?

Colour is not a factor that makes a difference to movement.

Thinking like a Scientist - Making a Prediction

We have already learned that scientists, like us, **observe** the world around them. Scientists always make very careful observations by paying close attention using their **senses.** We can also do this when we are learning about our world.

When we observe or investigate something, we often have some ideas about what we think might happen. We call this **predicting.** It means **making your best guess** based on what you already know.

Imagine you throw a ball up in the air in front of you. What do you **predict** the ball will do? What is your best guess?

____

Did you predict it will drop to the ground and probably will bounce up and down a bit until it stops?

____

TARGETING SCIENCE FOUNDATION © PASCAL PRESS ISBN: 9781925726497

Thinking like a Scientist - Asking Questions

Scientists are always **asking questions** in their work. This is how they discover new things.

We can learn to ask science questions too.

Why do super balls bounce so high?

How does the air get in the ball?

What would happen if we put water in the ball instead?

Why do some balls have air in them?

Look at the objects around the room. Think of some science questions you could ask to find out more about things you see. Perhaps search some of the questions on the internet with adult help.

Investigating Object Shapes

Some objects have a regular shape we can easily describe, like these objects. What shapes can you see?

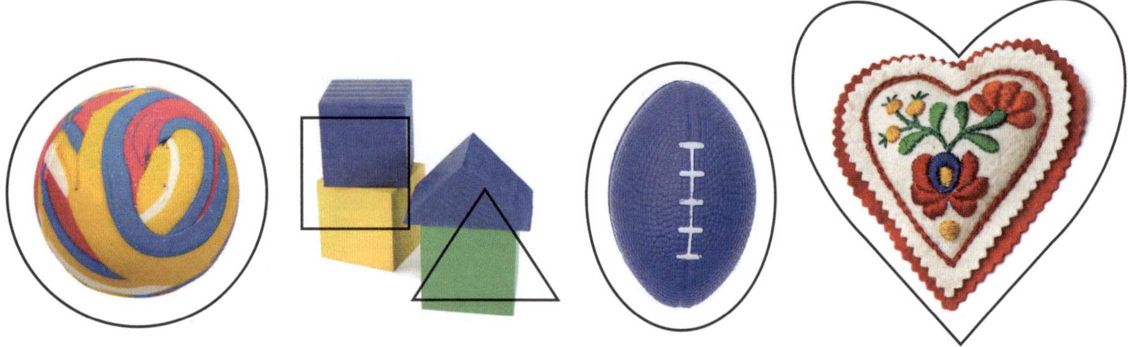

But many objects are shaped differently. You already know lots of object shapes and because of this you know how to use them.

Look at the shapes below. Circle the one you would use to cut some paper to wrap a present.

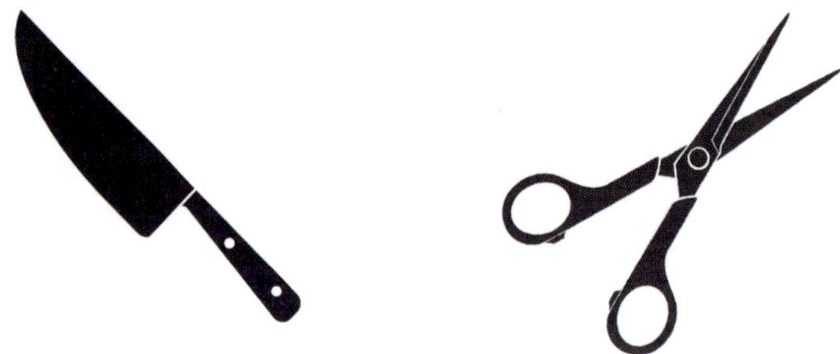

Circle the one you would use to join 2 pieces of paper together.

None of these objects have regular shapes like a circle or square, but we can still identify them by looking at their shape.

TARGETING SCIENCE FOUNDATION © PASCAL PRESS ISBN: 9781925726497

Investigating Object Shape and Movement

You will need 8 objects of different shapes and a way to make a ramp (see pictures).

OUR INVESTIGATION QUESTION IS: How does the shape of an object affect how it moves when it's pushed gently down a slope?

Collect 8 objects that are different shapes including a ball, a block, a tube, an egg-shape, a playing card, a sports shoe and a toy car or truck.

Use a toy slide or set up a ramp on a chair. This could be a piece of stiff cardboard, a length of wood or any other smooth stiff object.

You may need to secure the ramp with tape or string. Place a towel to stop the objects moving too far.

Investigating Object Shape and Movement

Now you are going to make some science observations. It is fun to release the objects down the ramp, but each time you do try to make careful observations like a scientist.

How does each object move down the ramp? Send all the objects 2 or 3 times down the ramp and observe closely. Repeat any objects you need to check again.

Share your ideas about these science questions:

1. What have you observed?
2. What words describe how the different shaped objects move down the ramp?
3. Do all the objects go down the ramp in a straight line?
4. Can you sort the objects into groups according to how they move?
5. Why did the toy car or truck roll?
6. Why did the block slide?
7. How does shape affect movement?

Take a photo of the groups you sorted to remember your thinking.

 TARGETING SCIENCE FOUNDATION © PASCAL PRESS ISBN: 9781925726497

Observing Toy Shape and Movement

https://clickv.ie/w/bzgx

Use this QR code to access a video on this topic.

Toys are one group of objects. The shape of a toy affects how it moves. You can probably guess how a toy might move from just looking at its shape. Look at the pictures below and work out what the movement word would be. Trace over the lines to show the movement.

Observing Toy Shape and Movement

Look at the toys below and write the word that best matches their movement.

roll bounce spin

____

____

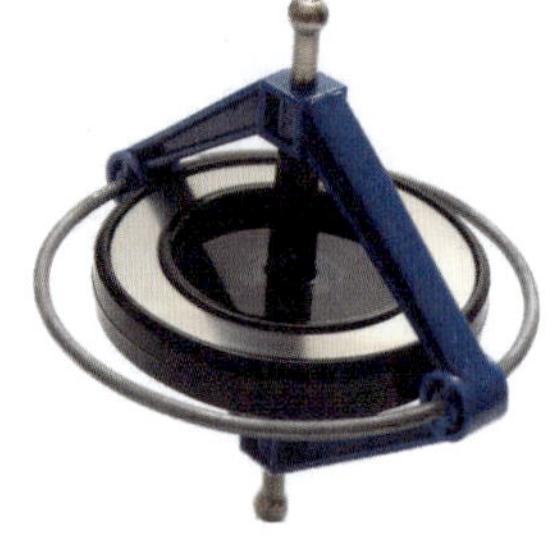

____

____

____

How did you know this is the way these toys would move?

Sorting Toys by Movement

Collect 10-12 toys that can move or have moving parts.

Observe each one and identify how it can move. **Sort** them into different **groups** depending on this movement and tell someone about the groups. Take a photo to remember your thinking.

Re-sort them into other groups. Take photos of the new groups.

Do you have any toys that move in 2 or 3 ways like this helicopter or basketball?

roll and spin

bounce, roll and spin

Ask a science question about one of your toys.

Observing Balls

Balls are such a fun object to play with, but have you observed them like a scientist?

Balls are often round. But not always.

Balls come in many colours and sizes.

Balls are made of different **materials** like…

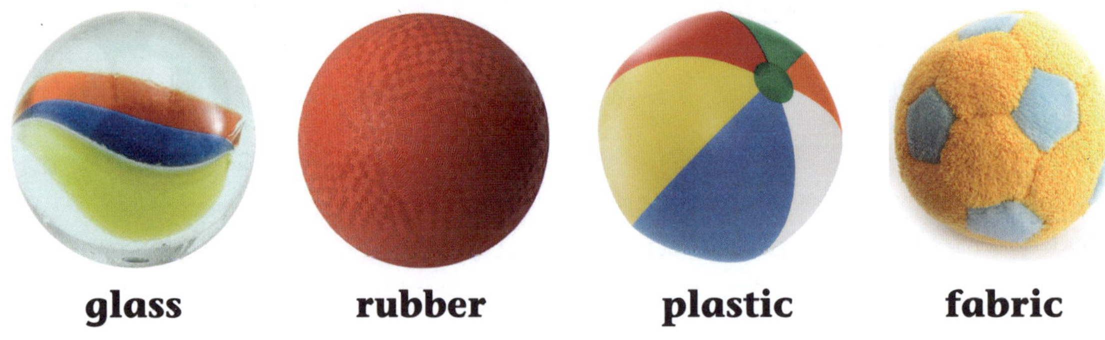

glass **rubber** **plastic** **fabric**

TARGETING SCIENCE FOUNDATION © PASCAL PRESS ISBN: 9781925726497

Investigating the Bounce of Balls

You will need: 5 **round** balls, including if possible, a 'super' ball, a marble, a plastic ball and a fabric ball.

When we investigate like a scientist we pose a question, make predictions, test, observe and record our observations. Then we think about what we have observed and learned.

Our science question today is: **why do some round balls bounce really well**?

Many balls can bounce. Collect 5 different **round** balls and observe them all closely.

Explain what you **observe**.

Now squeeze and feel all the balls. Explain what you observe. What do think the different balls are made from?

Talk with an adult to confirm what each ball is made from.

Movement of Objects

Investigating the Bounce of Balls

Remember our question for our science investigation is: Why do some round balls bounce really well?

To make it a fair investigation, you need to practise letting the balls go in the same way and observing them bounce. If you throw them onto the ground you may accidentally throw some harder and others more softly so it won't be a fair test.

Practise holding a ball in front of you with your arm stretched out like this and tipping, not throwing, the ball out of your hand.

Carefully observe it bounce. Practise this a few times so you can do it easily.

Look at the 5 balls you have for the investigation. **Predict** or make a guess about which ball you think will be the best at bouncing. **Predict** which will be the worst.

TARGETING SCIENCE FOUNDATION © PASCAL PRESS ISBN: 9781925726497

Investigating the Bounce of Balls

Draw the ball you predict will bounce best in the top row of the table below. Draw the ball you predict to be the worst bouncer in the bottom row of the table. Draw the other balls in the other rows.

Now test each of the balls by dropping them from your hand and observe. Retest them and tick ✔ the number of stars to show how each bounced.

Drawings of balls	How did it bounce? (a really good bouncer gets 5☆)
I predict this ball will bounce best	☆☆☆☆☆
	☆☆☆☆☆
	☆☆☆☆☆
	☆☆☆☆☆
I predict this ball will bounce the worst	☆☆☆☆☆

Investigating the Bounce of Balls

1. Did any of the balls surprise you with how they bounced? Why?
2. Did you predict which ball would be the best bouncer?
3. Were there other balls that bounced just as well?
4. Did you predict the ball that was the worst bouncer?
5. Does the colour of a ball affect how it bounces?
6. Was the biggest ball the best bouncer?
7. Why do you think some balls are good bouncers and others are not?

Note: Size is a common misconception children may have about the bounciness of balls. The material a ball is made from is the main factor that influences how bouncy a ball is. There are other factors like shape, how much air (if it's an inflated ball), where you bounce it, hardness or softness of the rubber and surface covering. You may like to explore some of these variables.

TARGETING SCIENCE FOUNDATION © PASCAL PRESS ISBN: 9781925726497

Exploring Traditional Ball Toys

For thousands of years, First Nations children have enjoyed ball games just like we do. Sometimes they would find ball-shaped objects and other times adults and teenagers would help them find materials to make them. Ball games would help them learn important skills they would need to know when they became adults.

Stones found near a river

Ball made from plant fibres from Cairns area.

Ball made by Calcadoon and Western tribes of the Georgina River, Channel Country QLD, It is made from animal (bullock) bone, gypsum, camel dung and pigment and spun on a paperbark mat with sand.

Balls made from plaited Pandanus leaves by the Meriam people of Mer Island in the Torres Strait.

First Nations children were always careful to play safely with stones and sticks so they didn't hurt themselves or others.

How do you think you can play safely with sticks and stones?

Exploring Traditional Ball Toys

A favourite game of the First Nations children in the Numinbah Valley area of South Queensland was to roll pebbles down long tubes of bark. They would let them go at the same time and dash to the bottom of the tube.

(main image Narinda Sandry, pebble added)

The first pebble out was the winner.

The children would take great care in choosing a pebble to race.

What sort of pebble would you choose to play this game? Name 3 things you would look for.

____

____

____

TARGETING SCIENCE FOUNDATION © PASCAL PRESS ISBN: 9781925726497

Investigating Rolling Balls

You will need: plasticine or playdough, a ramp setup and a ruler.

You are going to find and make some small balls to do a science investigation.

Our science question is: How does shape affect rolling balls?

Part A 

1. Use some plasticine or playdough to make 2 different balls that you predict will roll well down a ramp.
2. Find 2 different stones or pebbles that you predict will roll.

3. Set up a ramp outside or inside. You will use the ruler to hold the balls at the top of the ramp until you are ready to release them. Don't make the ramp too steep. This way you have time to observe closely.

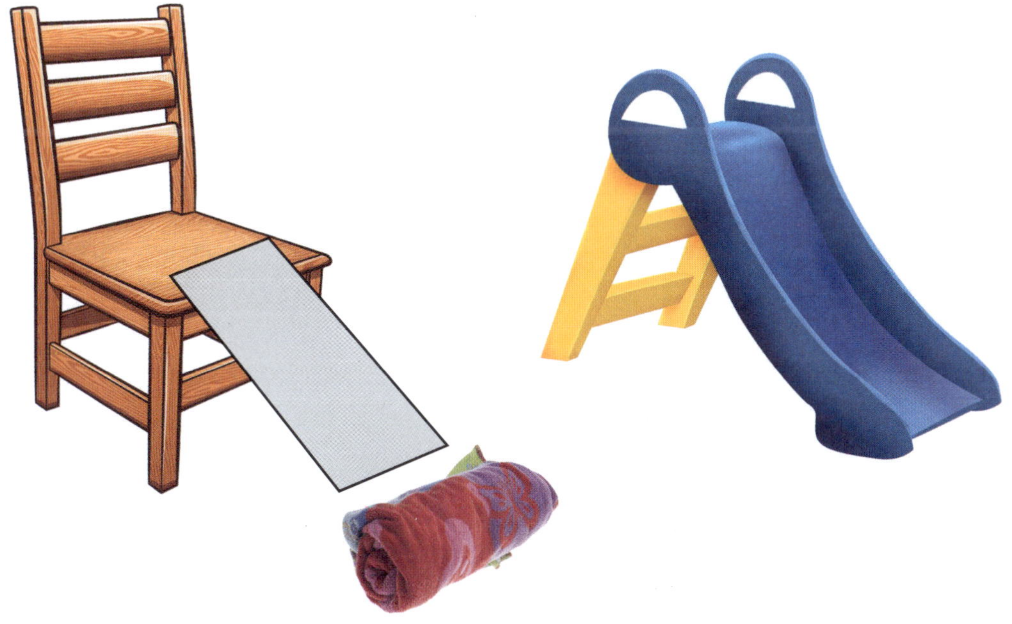

Investigating Rolling Balls

4. Place your 4 objects in front of you and observe them carefully like a scientist.

5. Draw them in the boxes below.

1.	2.
3.	4.

6. Now **predict** which of the 4 will roll the fastest down the ramp. Put a line under the drawing of it.

7. Release the balls with the ruler at the same time and observe the way they roll, as well as which one wins.

8. **Test** roll them 5 or 6 times down the ramp and observe closely. You may like to put a dot or tally in the boxes above every time one wins.

9. You can try just rolling 2 at a time to compare different ones.

TARGETING SCIENCE FOUNDATION © PASCAL PRESS ISBN: 9781925726497

Thinking about your learning so far:

Which one rolled fastest most of the time?
Number __________

Why did this one roll so well? ____________________

Did you predict this would be the fastest roller?
Yes   no

It doesn't matter if you picked a different one. Scientists do this too.

Did they all roll straight down the ramp? If not, why do you think they didn't?

____

Part B. 

10. Now take the 2 ball shapes you made yourself. We are going to test your thinking.

11. What do you think will happen if you make one of the ball shapes into a box shape?

a. Make the box shape and predict what will happen.

b. Test the new shape on the ramp and observe what happens.

c. Were your predictions correct? Yes No

12. Now take the box shape and ask you own science question:

What will happen if ...

Change the box shape into another shape and ...

- ✔ PREDICT
- ✔ TEST AND OBSERVE
- ✔ COMPARE with what you predicted

Write a sentence about shape and rolling:

___

___

___

___

TARGETING SCIENCE FOUNDATION © PASCAL PRESS ISBN: 9781925726497

Think about what happened when you made the box shaped 'ball' and tried to roll it down the ramp.

People used this science knowledge to invent wheels. Wheels made people's lives so much easier. In the beginning they made them from pieces of tree trunks and stone. They were very heavy!

Stone wheels

Wooden log wheels

Images source: Wikimedia commons

Wheels are shaped to help them move easily but also stay in place.

Wheels are designed to spin on an axle. The axle is usually in the middle and holds the wheel on to the frame. What do you think would happen if wheels were made square?

Using our Science Learning - Wheels

Circle the things that have wheels.
Can you point to some axles?

TARGETING SCIENCE FOUNDATION © PASCAL PRESS ISBN: 9781925726497

Using our Science Learning - Wheels

Draw a line to match.
How do the wheels help each object move?

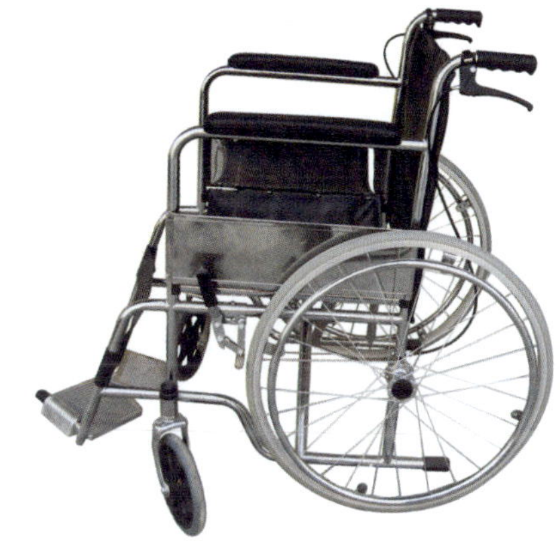

Wheels Help Us Work

Draw a line to match.

TARGETING SCIENCE FOUNDATION © PASCAL PRESS ISBN: 9781925726497

Trace the wheels.

Wheels Go Round

Wheels do the work
From here to there,
On a toy
Or on a chair.

Wheels do the work
From here to there
All through the town.

Talk with Your Child

Explain to your child that wheels help make work easier for people. Think about the things you use that have wheels. Talk about how they help make work easier for your family.

TARGETING SCIENCE FOUNDATION © PASCAL PRESS ISBN: 9781925726497

Draw a line to match.

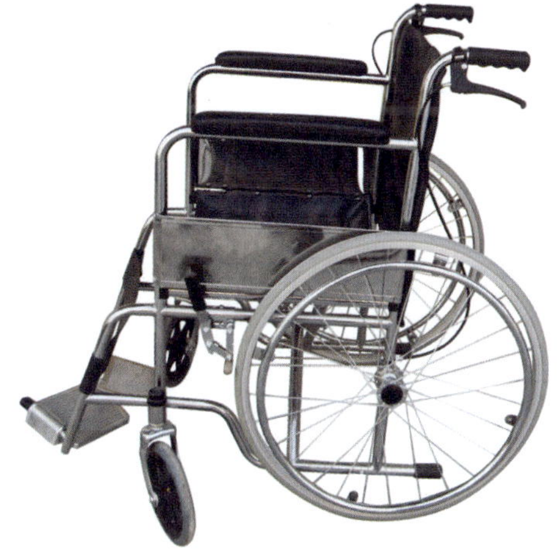

Objects in our World

In this science unit we are going to learn more about objects.

Do you remember…

Objects are things in our world
which can be seen, held or touched.

We use objects every day. Objects are not alive.

Objects can be different **colours**

Objects can be different **shapes**

Objects can be different **sizes**

Objects can be made from different **materials**

TARGETING SCIENCE FOUNDATION © PASCAL PRESS ISBN: 9781925726497

Observing Materials

https://clickv.ie/w/Gygx

Use this QR code to access a video on this topic.

The material an object is made from is very important. It affects what an object can do and the way we use it.

Look at the list below. Now find a real object made from each material. Look at it and feel the object, then draw it.

Material	Can be used to make	Draw something else you found made from this
wood		
plastic		
metal		
fabric		
glass		
paper and cardboard		

TARGETING SCIENCE FOUNDATION © PASCAL PRESS ISBN: 9781925726497

Observing Materials

You will need: Ask an adult to collect 15 safe things used in a kitchen, made from different materials, and put them in a bag you can't see through.

When you have the bag, put your hand in and feel 1 object at a time. DON'T PEEK!

BEFORE taking it out of the bag, try to:

Name the object…

Name the material…

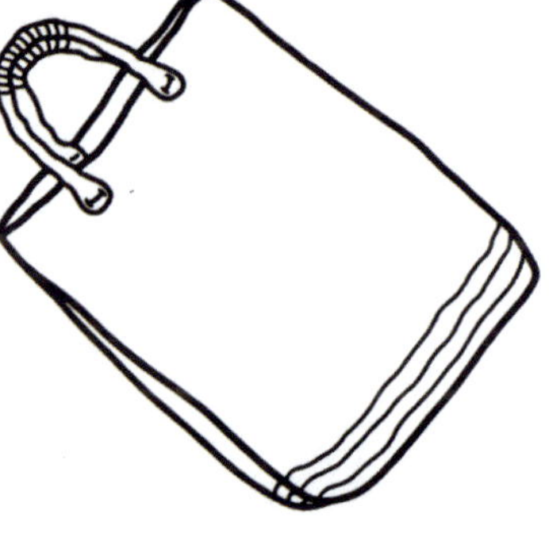

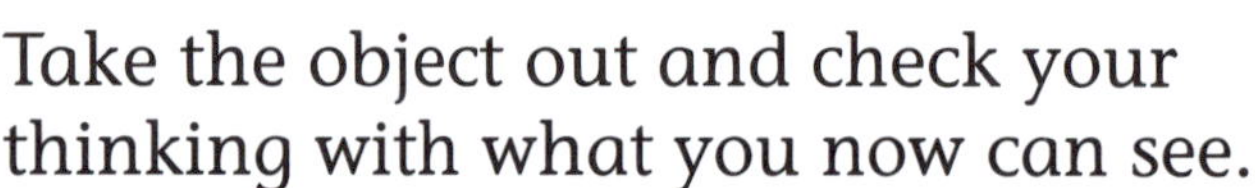

Take the object out and check your thinking with what you now can see.

Do this for all the objects.

Now sort and group the objects by their material.

Look at each material group and observe the different shapes, colours and how they feel.

Are the objects in each group the same or different somehow?

Take some photos to remember your learning.

It is wonderful that we have so many different materials in our world. Each **material** has special things about it. We call these things its **properties.**

Glass is see-through and strong, stiff, smooth and can be thin. It can be wiped clean. These are some **glass** properties.

Imagine if the lenses were made from cardboard!

Fabric is bendy or flexible so we can shape it around our bodies. It can be very soft and warm. It's strong enough to wash when it needs cleaning. It can be sewn into many shapes like blankets and clothes. These are some **fabric** properties.

Imagine if the blanket was made from metal!

Glass and Fabric have Properties

Tick which material, **glass** or **fabric**, you think would have the best properties for each object.

Object	Better made from **glass**? Why? 	Better made from **fabric**? Why? 

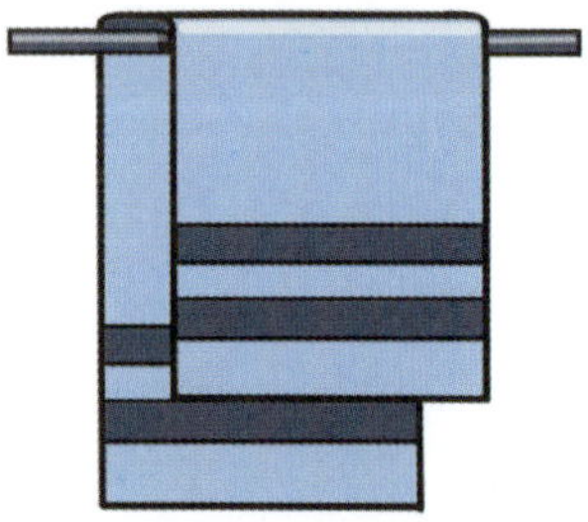		

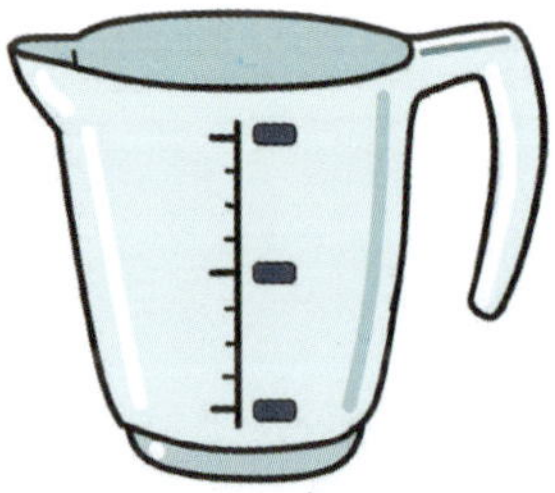		
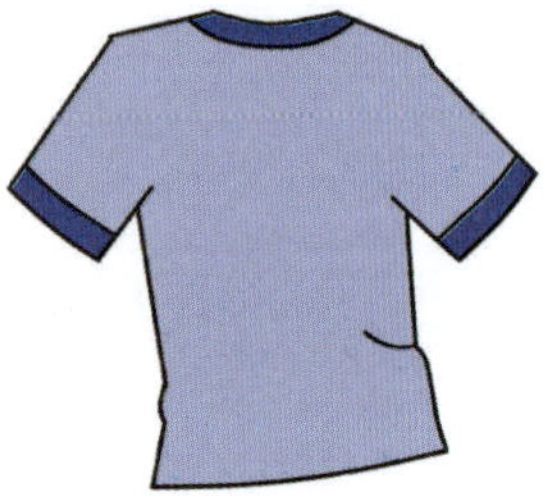		

TARGETING SCIENCE FOUNDATION © PASCAL PRESS ISBN: 9781925726497

Look at my swim toys, do you see? What makes them great is their properties!

Plastic is very waterproof so it can be put into water and not fall apart. It can also hold water like a cup does. It can be stiff or flexible (bendy) which makes it great for many different uses. Plastic is often smooth. It can be coloured or clear and made into any shape you can imagine, even shapes that hold air! These are some plastic properties.

Imagine if my water toys were made from paper!

Look at the toilet paper, do you see? What makes it great is its properties!

Paper and cardboard are very useful materials. Paper can be made soft for the toilet but stiff for drawing on. Very stiff and thick layers of paper become cardboard which can be strong for boxes. Paper and cardboard go soggy if wet. Paper can be folded and comes in many colours. Paper and cardboard make books.

Imagine if toilet paper were made from glass or metal or wood!

Investigating Waterproof Materials

When we say an object is waterproof, we mean the material it is made from does not allow water to go through it. Sometimes this is a very important property. A raincoat and gum boots would be useless if they allowed water through to your clothes.

You are going to investigate the property of being **waterproof**. You will need:

- a bowl of water
- 6-7 small objects made from different materials (or pieces of objects/ materials) e.g. piece of paper, unpainted wood item, rubber ball, plastic toy/cup, metal toy/ spoon, fabric object, drinking glass, ceramic object
- a towel to put things on after you have tested them

Remember when we investigate like a scientist we pose a question, make predictions, test, observe and record our observations. Then we think about what we have observed and learned.

You will need to set up your bowl of water, towel and materials in a safe place and your recording page nearby.

Materials

TARGETING SCIENCE FOUNDATION © PASCAL PRESS ISBN: 9781925726497

Investigating Waterproof Materials

Question: What materials have the property of being waterproof?

Object (write or draw)	I predict waterproof	What happened?	✔ I predicted correctly. or I learned a new thing

What have you learned?

Wood and Metal have Properties

Look at our trainset, do you see? What makes it great is its properties!

Wood is hard, stiff and strong and can be carved into many shapes. It can be made smooth, and it can be painted or left a natural colour. It smells and looks beautiful. It soaks up water but will dry out again. These are some wood properties.

Would the trainset be good made from glass?

Look at our barbeque and tools, do you see? What makes them great is their properties!

Metal is very strong and often stiff. It can be made very hot but it will not fall apart or burn up. It can be made into many shapes including for cooking utensils, tools, machines and cars. It is waterproof and lasts a very long time. It is often smooth. These are some of the properties of **metal**.

There would be no trains or train tracks without metal, and no buses, cars or aeroplanes. Oh and no barbeques!

Hold a metal spoon and name some properties.

TARGETING SCIENCE FOUNDATION © PASCAL PRESS ISBN: 9781925726497

Circle the one in each row that is made from a different material

Important Properties

Sometimes certain properties are more important than others for an object.

For traffic lights, **colour** is one important property.

Sometimes **hardness** is important.

A table needs to be **hard.**

But a pillow needs to be **soft.**

Flexibility means how stretchy and bendy something is. If something isn't flexible we say it is stiff or **rigid.**

Things like rubber bands and hair ties need to be **flexible**.

But we need the walls of our buildings to be **rigid and stiff**.

TARGETING SCIENCE FOUNDATION © PASCAL PRESS ISBN: 9781925726497

Look at these objects. What are they made from? Why are they made from this material?

Material: ______________________________

Why? ______________________________

____

____

Material: ______________________________

Why? ______________________________

____

____

Pose some science questions. For example: What if nails were made from fabric or glass?

Texture

Materials feel different to touch. We call this its **texture**. Sometimes texture is an important **property** for an object.

A **soft** and **fluffy** teddy bear feels good to cuddle. But would it feel the same if it was made from **hard** clay?

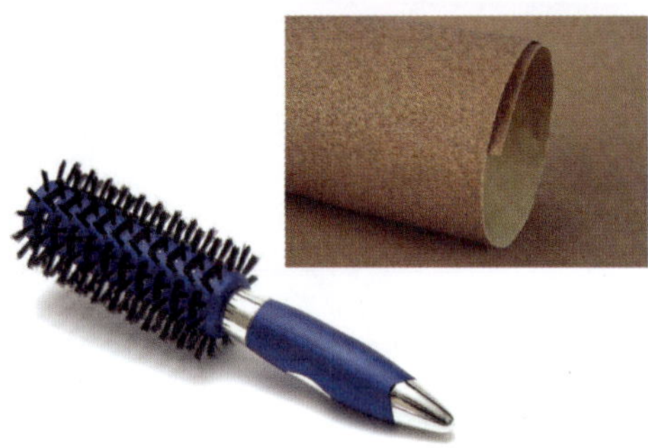

A hairbrush needs to feel **spikey** for it to work properly.

Sandpaper has to feel **rough** to sand the wood to make it very **smooth**.

A fridge needs to feel cold on the inside or the food will spoil.

A kettle would be broken if it didn't make the water **hot**.

And what would be the point of glue or tape if they didn't feel **sticky**!

Materials feel different to touch. We call this its **texture**.

 TARGETING SCIENCE FOUNDATION © PASCAL PRESS ISBN: 9781925726497

Draw a line to **soft** or **hard**.

soft

hard

Objects Made from one Material

The same type of object can sometimes come in different materials. Each material has its own special properties.

Find as many different balls as you can. Squeeze, feel and bounce them.

Now sort them into groups by the material they are made from.

Take a photo to remind you of your thinking.

Look at these balls. Name the material used. What is 1 property of each material used?

______________________ ______________________

______________________ ______________________

TARGETING SCIENCE FOUNDATION © PASCAL PRESS ISBN: 9781925726497

Many **objects** are made from just one type of **material,** but lots of objects are also **made from 2 or more materials**.

The materials used have different properties.

Why was metal used? Strong, smooth, hard, stiff. These are the properties.	**Why was fabric used?** Flexible, soft, strong, colourful, pieces can be sewn together.
Why was rubber used? Waterproof, flexible, grips the ground, strong.	**Why was plastic used?** Strong, hard, waterproof, sticks tight to the fabric.

All of these materials and their properties make this shoe work. Would a shoe be any good if it were made only from metal?

Objects and their Parts

Look at this bicycle. Think about all the different materials used. Name as many as you can.

Why are the tyres made from rubber?

____

____

Why is the frame made from metal?

____

____

The materials are chosen because of their properties.

 TARGETING SCIENCE FOUNDATION © PASCAL PRESS ISBN: 9781925726497

Grouping Objects and their Materials

https://clickv.ie/w/D0gx

Use this QR code to access a video on this topic.

You will need: 10-12 toys, some made from 1 material and others 2 or more materials.

Toys can be made from many different materials. Observe the materials of some different toys, like a scientist. This means looking carefully and feeling but not playing.

Try to identify what materials have been used to make each toy.

Point to a toy you have that is

- ✔ only **one colour** and one that has **many colours**
- ✔ one that is **hard** and one that is **soft**
- ✔ one that is **bendy (flexible)** and one that is **stiff (rigid)**

Sort the toys into 2 groups - the toys made from just 1 material and the toys made from 2 or more materials.

Materials

Grouping Objects and their Materials

Choose one of the toys made from **more than one** material.

✔ Name 2 of the materials that have been used for different parts.

____________________ ____________________

✔ Choose one material and describe some of its properties.

____

____

✔ Are the properties of this material important for the toy?
Yes No

✔ Look at these toys. Put a dot on the ones you could take in the bath.

Why did you choose these?

____

____

____

TARGETING SCIENCE FOUNDATION © PASCAL PRESS ISBN: 9781925726497

First Nations Australians' Use of Materials

First Nations Peoples have lived in Australia for over 60,000 years. For most of these years, there were no shops or factories to buy things from. But they were very clever and found the materials they needed from the plants, animals and rocks around them. They used these to make things to help them survive.

First Nations Australians knew that flint stone had the best properties for shaping into tools for hunting and cooking.

They would hit it and scrape it on other rocks to make very sharp pieces.

Then they would lace these to pieces of wood using vines that had flexible and strong properties. They would also use resin or tree sap because of its properties to help hold the vines in place on tools like these.

First Nations Australians' Use of Materials

First Nations Australians also knew which leaves had the best properties for weaving into baskets and mats.

They knew where to find rocks with the best properties for grinding into powder for making into paint.

First Nations Australians also understand that Stringybark and Woollybutt trees have the best properties for carving and making into didgeridoos.

The properties of the materials they chose were very important.

Observe some natural materials in a garden or park. Name their properties.

TARGETING SCIENCE FOUNDATION © PASCAL PRESS ISBN: 9781925726497

Using Special Tools for Observing Materials

When scientists want to look at something very small or they want to zoom in and look at what something is made from, they use special tools. They might use a magnifying glass or a microscope.

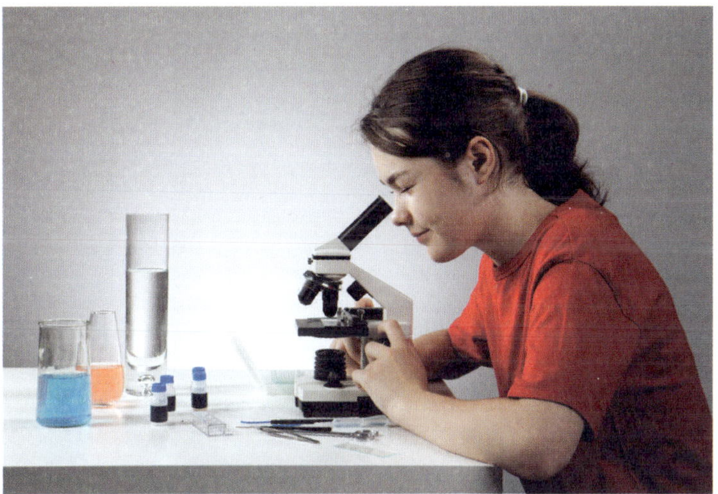

Scientists are always looking to understand and make materials with better properties. Here is a close-up photo of a waterproof fabric. A scientist can look at this and see how well the fabric stops the water from soaking in.

Do you think it is working well?

Using Special Tools for Observing Materials

If you have a magnifying glass or a digital camera you can set to macro, take a look at some different materials yourself.

If not, here are some to show you the amazing things you might see.

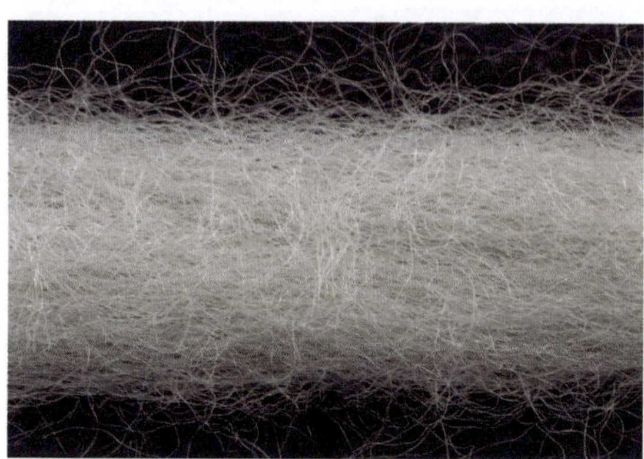

Wool thread

Slice of wood

Do you have any Velcro shoes? Did you know that a scientist named George de Mestral found that certain prickles stuck really well to his clothes when he went hiking. He took some to his laboratory and looked at the prickles under a microscope and saw they had tiny little hooks.

He made something out of plastic with the same properties and invented Velcro!

TARGETING SCIENCE FOUNDATION © PASCAL PRESS ISBN: 9781925726497

Sing this science chant to your child to the tune of "I'm a Little Teapot".

I'm a little magnet.

Watch and feel—

I stick to iron

And I stick to steel.

When I'm near some metals,

I will pull.

The attraction is powerful!

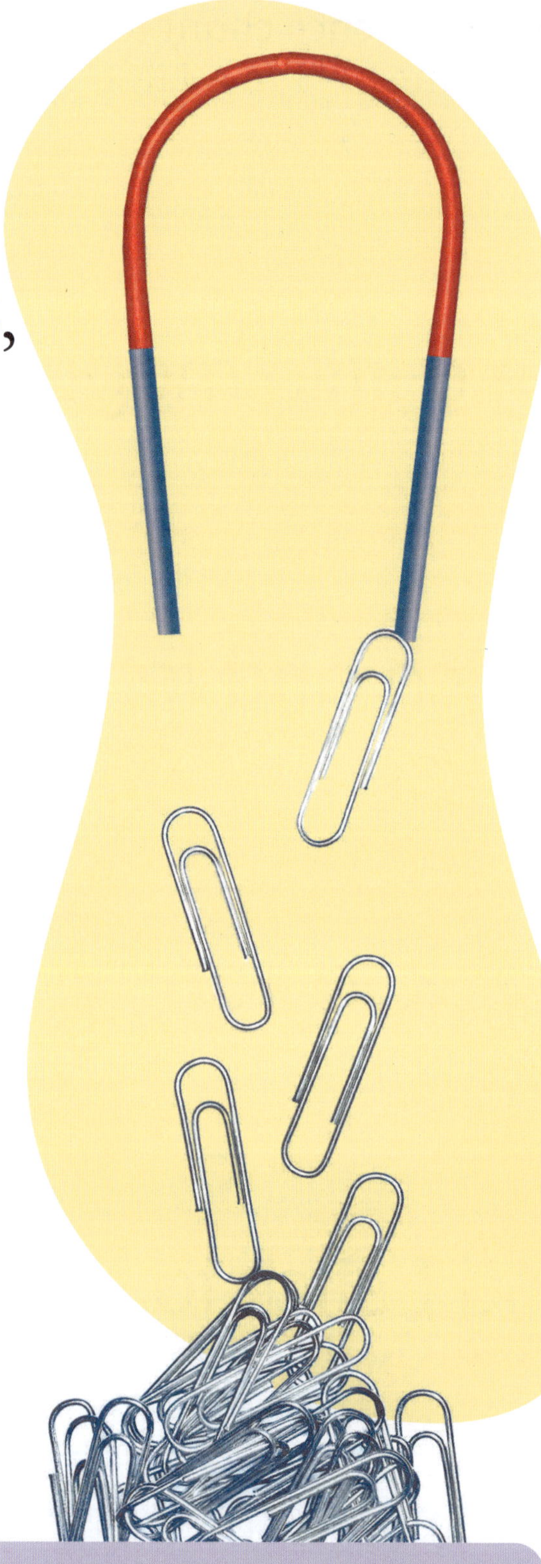

Talk with Your Child

Obtain some magnets and allow children to explore. Explain to your child that a magnet is a kind of metal that pulls on certain things. Tell your child that many things that are metal will stick to a magnet, but things that are not metal such as crayons, socks, and cookies will not stick to a magnet.

TARGETING SCIENCE FOUNDATION © PASCAL PRESS ISBN: 9781925726497

Magnetic-a Special Property

A special property that many but not all metals have is being able to stick to a magnet.

Draw a line to identify which objects would be magnetic and which would not.

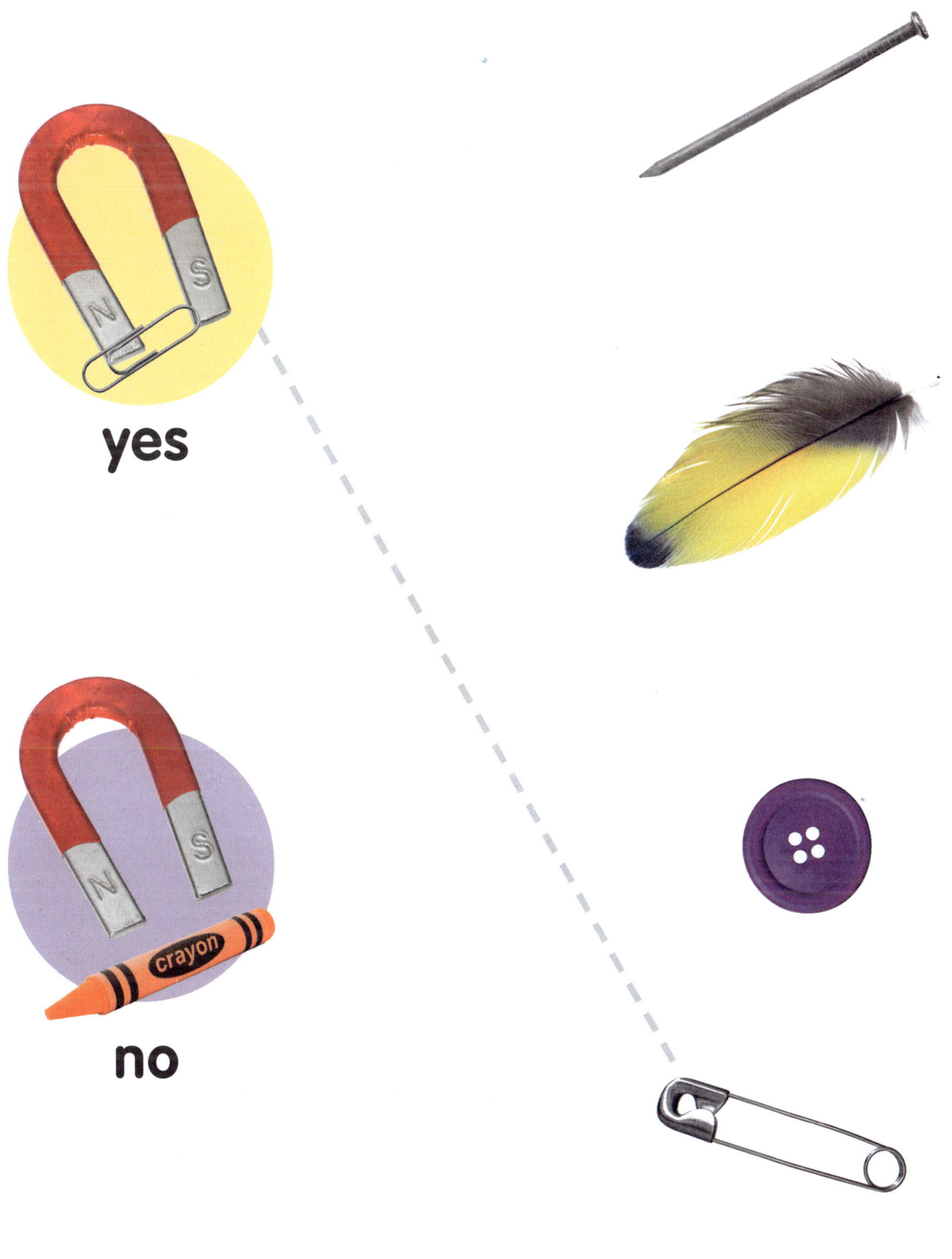

What Will Stick?

Circle the thing that a magnet will stick to.

1

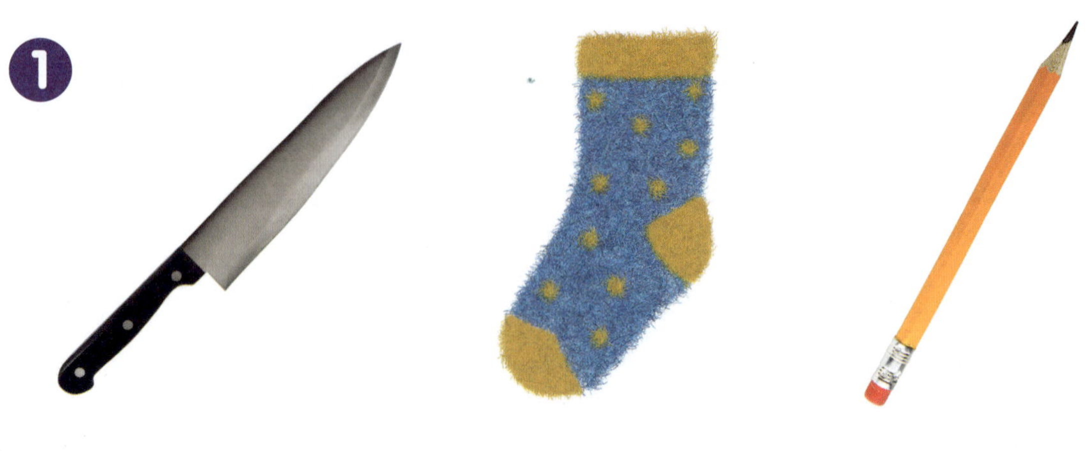

2

3

Materials

TARGETING SCIENCE FOUNDATION © PASCAL PRESS ISBN: 9781925726497

Page 6

(top to bottom) lizard; emu; brolga; kangaroo

Page 7

(top to bottom) fish; snake; bat; lizard

Page 11

snake yes; tortoise yes; cat no; bird no

Page 12

1 parrot; 2 chick; 3 squirrel; 4 rabbit. They all have feathers

Page 13

feathers: peacock, parrot, owl; fur: cat, rabbit, bear

keep warm, enable to fly, protect from sunburn and insect bites

Page 16

dog: paws; duck: webbed feet; eagle: claws; horse: horse shoes

Page 18

paw: dog; fishtail: goldfish; hooves: horse; wing: bird

Page 28

leaves: lettuces; roots: carrots, onions, potatoes; fruit: fruit; seeds: beans

Page 29

fruit: e.g. grapes, apples, berries; leaves: lettuce, spinach; flower: cauliflower, broccoli; stem: asparagus, celery; seeds: passionfruit; root: radish, onions, carrots

Page 30

spinach: leaf; carrot: root; orange: fruit; peas: seed

Page 31

1 seed; 2 tomato; 3 lettuce

Page 38

ball, Lego house, blocks, stuffed toy, pencil, shoe

Page 40

size: 3rd ladder is smaller; shape: 2nd mug has handle; weight: 1st image has lots of bricks; material: 3rd bowl is metal; made to move: 2nd object is a stationary slide

Page 48

skates: roll; ball: bounce; top: spin; bike: roll; trampoline: bounce

Page 56

texture, size, shape, composition, weight

Page 62

pram, roller blades, pizza cutter, wheelbarrow

Page 63

backpack: luggage; chair: wheelchair; basket: trolley

Page 64

fruit: trolley; pizza: pizza cutter; soil: wheelbarrow

Page 67

shoes: roller blades; chair: wheelchair; bag: luggage

Page 72

glass: lightbulb so light can shine through; jug so can hold and see liquid;

fabric: towel so will be soft and absorbent; sofa so comfortable and won't break; shirt so can be worn

Page 77

rows 1-4: 3rd image; row 5: 1st image

Page 79

nail: metal - hard, rigid, sharp, can be shaped

sock: fabric - soft, warm, smooth, can be shaped

Page 81

soft: cushions, teddy; hard: walnut, bucket

Page 82

fabric: soft; rubber: bouncy; glass: hard; plastic: flexible

Page 83

No, it would not be comfortable. It would be cold (or hot on hot days) and rigid.

Page 84

Rubber is flexible and strong and can take impact without breaking. Metal can be shaped, and it is rigid and strong and can stop the bike frame from warping.

Page 92

yes: nail, safety pin; no: feather, button

Page 94

1 knife; 2 scissors; 3 pushpin